VISITING EDEN

VISITING EDEN

the public gardens of northern california

Photographs by Melba Levick ✺ *Text by Joan Chatfield-Taylor*

CHRONICLE BOOKS

SAN FRANCISCO

Book design by Adelaida Mejia Design, San Francisco
Editing by Carey Charlesworth.

Printed in Japan.

Library of Congress Cataloging-in-Publication Data

Chatfield–Taylor. Joan.
Visiting eden: the public gardens of northern California / Photographs by Melba Levick;
Text by Joan Chatfield–Taylor.
p. cm.
Includes index.
ISBN 0–8118–0107–1.
1. Gardens—California, Northern. 2. Gardens—California, Northern—Pictorial Works. 3. Botanical
Gardens—California, Northern. 4. Botanical Gardens—California, Northern—Pictorial Works. I. Levick.
Melba. II. Title.
SB466.U65C23 1993
92–17134
712. 5'09794—dc20
CIP

Distributed in Canada by Raincoast Books,
112 East Third Avenue, Vancouver, B.C. V5T 1C8

10 9 8 7 6 5 4 3 2 1

Chronicle Books
275 Fifth Street
San Francisco, CA 94103

I WOULD LIKE TO ACKNOWLEDGE WITH GRATITUDE all the garden designers and caretakers without whose wonderful talents and efforts these "edens" would not exist for us to enjoy. I am privileged to work with Chronicle Books, whose excellent taste and quality I can always depend on, and I would like especially to thank my editor, Nion McEvoy, for making this extremely pleasurable and rewarding experience possible. I would also like to thank my dear friend Teresa Cathutrecasas for so generously sharing her home with me while I worked so happily in these gardens. Lastly, I want to thank Joan Chatfield-Taylor for the pleasure of working with her.

—ML

I WOULD LIKE TO EXPRESS MY DEEPEST APPRECIATION to all the gardeners and garden-lovers who design and nurture the magical spaces described in this book and who took time to discuss their work with me. I give special thanks to copy-editor Carey Charlesworth for her meticulous attention to detail and the many others at Chronicle Books who made this book a reality. In addition, my thanks go to Melba Levick, whose unflagging enthusiasm was a constant inspiration.

—JCT

CONTENTS

UNIVERSITY OF CALIFORNIA SANTA CRUZ ARBORETUM

Santa Cruz, California

The University of California's Santa Cruz campus has a reputation as the most individualistic and idiosyncratic school in the system. Its Oxford-inspired colleges are part of the reason, but so is a lifestyle that in many ways perpetuates the iconoclastic spirit of the 1960s. This is a campus that chose the banana slug as its mascot and where the environment is both a major and a religion.

The university's arboretum is perfectly in tune with all this. Although it's the youngest and poorest of the University of California's arboreta and botanical gardens, it's not a poor imitation of its better-established cousins. The Santa Cruz arboretum is known as a repository of unusual exotic plants, especially those from Australia, New Zealand, and South Africa.

Ray Collett, a professor at UC Santa Cruz and director of the arboretum, explains, "When we got started, we were quite aware that California had a poor selection of Mediterranean climate plants, particularly Australian plants, that would do well here. Fortunately, Australians are very proud of their native plants, so we got into the loop there and were able to avail ourselves of a treasure trove of native plants. Thanks to Qantas' direct flights, we can get even the trickiest and most finicky plants here just a few hours after they're picked."

Once across the Pacific, the new specimens usually thrive on this frost-free, one-hundred-thirty-five-acre site, warmed by breezes from the ocean and Monterey Bay.

The Australian section is the most mature part of the garden. The curving pathways are flanked by tall trees and massive shrubs, predominantly eucalyptus, grevilleas, and banksias, each represented by an astonishingly varied array of species. The grevilleas, for instance, include *Grevillea aquifolium*, with leaves resembling Christmas holly, *Grevillea lavandulacea*, whose spurred pink flowers suggest columbines, and *Grevillea gaudichaudii*, a deep green ground cover with red flowers. The banksias are equally odd and strange; note *Banksia laevigata*, with its flowers like balls of fur, *Banksia oreophila*, with its hairy pods, and the narrow-leaved hairpin banksia.

Other eye-catching exotics in this section include the graceful *Eucalyptus pauciflora*, with bark and leaves that look as if they've been dipped in silver paint, *Melaleuca thymfolia*, whose flowers resemble puffs of fine white lace, and *Acacia baileyana purpurea*, with blue-to-lavender pods that give the whole tree a purplish cast.

Home gardeners should not start planning their spring planting around these items. Many of these natural wonders are not available commercially because the arboretum is still testing them, a process that takes as long as twenty years. Because the Australian species are growing here without their usual complement of pests, they often grow to unexpected sizes.

Now that Australia is well established, the arboretum is concentrating on the New Zealand section. UC Santa Cruz also boasts one of the broadest collections of New Zealand species in the United States. Although this section is relatively new, Collett expects that the plants will eventually grow to form a temperate rain forest, in dramatic contrast to the dry-climate plants of Australia.

"The flora of the two countries are entirely different. Australia is bright colors, perhaps because the plants are competing for attention from the insects during the short wet season. In New Zealand, where there's no shortage of water, almost all the flowers are white, with astonishing foliage texture."

The contrast underlines one of the basic themes of the Santa Cruz arboretum, the shifting of tectonic plates that split the southern continent called Gondwanaland into the separate landmasses of Africa, Australia, New Zealand, and India, each with its own biological identity.

One of the most colorful areas in the arboretum represents another of these landmasses, Africa. The stars of the South African section are the hundreds of different proteas, many with enormous, brilliantly hued flower heads. California gardeners had considered proteas almost impossible to grow until the Santa Cruz arboretum discovered just what kind of benign neglect suited these spectacular plants. South African companion plants, particularly the leucospermums, add even more color to this dramatic area.

Visitors should realize that this is not a long-established, well-financed institution. As garden manager Brett Hall puts it, "We're young and we're poor." This means a lack of signage, other than the hard-to-read metal tags on some plants, so visitors are advised to take advantage of the frequent tours given by volunteer docents. Furthermore, the tiny staff, even when augmented by students, is unable to do all the weeding.

"Maybe it's a good thing we don't get all the weeding done. It gives a feeling of authenticity," says Brett Hall. The exotics, popping out of a carpet of grasses, weeds, and wildflowers, seem to be growing naturally in the wild. An abundance of animals, such as lizards, squirrels, deer, and gopher snakes, adds to the sense of being in a natural environment. Hummingbirds like it here, as do monarch butterflies, who have taken to wintering in the eucalyptus grove.

❀ UNIVERSITY OF CALIFORNIA AT SANTA CRUZ ARBORETUM
University of California at Santa Cruz Santa Cruz, CA 95064 408/427-2998. Daily, 9 to 5. Donation requested. In Santa Cruz, follow signs toward the university. At the intersection of Bay and High streets, turn left on High street; take it north ½ mile.

Paths wind informally
through the arboretum's
large collection
of unusual trees and shrubs
from Australia.

VILLA MONTALVO

Saratoga, California

The sight of Villa Montalvo on a summer evening, when concertgoers promenade on sloping lawns dappled with shade, is a vision of another, more gracious era than our own. Montalvo, one of the great estates built on the San Francisco Peninsula in the early years of this century, was the creation of James Duval Phelan, a Renaissance man who was one of the busiest figures in San Francisco history.

Born in San Francisco in 1861, Phelan was the son of an Irishman who brought three boatloads of merchandise to sell to the gold miners in 1849 and then had the foresight to buy up large quantities of California real estate. The second-generation Phelan, besides managing the family's considerable fortunes, served as San Francisco's mayor from 1897 to 1902, fighting corruption and initiating plans for the Hetch Hetchy water system that serves the city to this day. In 1915, he became the state's first popularly elected United States senator.

His interests were more than political. At various times in his life, Phelan served as president of the Bohemian Club, chairman of the Mid-Winter Fair (in 1894), president of the San Francisco Art Association, chairman of relief efforts following the 1906 earthquake, trustee of the San Francisco Public Library, and head of the Recreation and Parks Commission.

Small wonder that Phelan felt the need for a place in the country. In 1911 he purchased land in the foothills of the Santa Clara

Senator Phelan's deep affection
for Italy is expressed
at Villa Montalvo
by numerous Italianate statues
and a symmetrical layout
reminiscent of
Mediterranean gardens.

Valley, near the small town of Saratoga. There he built his dream house and filled it with Oriental rugs and furnishings that he purchased in Europe.

From the veranda, Phelan looked down a wide sweep of lawn towards the Temple of Love, a folly composed of twelve Ionic columns. Besides providing an impressive view of the Santa Clara Valley, then a rural landscape of woods and orchards, the structure expressed Phelan's affection for the city of Rome and for all things classical.

In front of the Temple of Love is a rigorously symmetrical garden with statuary and tall cypresses to emphasize its Italianate style. The central beds, outlined in different types of boxwood, contain a collection of citrus trees, including several varieties of lemons, oranges, tangerines, and mandarins. Other plantings in this level area include Chinese holly (*Ilex cornuta*), smoke tree (*Cotinus coggygria*), rose of Sharon (*Hibiscus syriacus*), and silk tree (*Albizia julibrissin*), with its canopy of pink flowers.

On either side of the lawn, Phelan and his head gardener, George Doeltz, planted a large variety of trees, including many rare and unusual varieties. The collection is so extensive that Villa Montalvo is considered an arboretum, although it does not have a formal arborist and some of its specimens are not labeled.

A walk along the pathways that skirt the lawn is a glimpse of forests around the world, because Phelan chose trees from almost every continent. A few examples of his far-reaching taste include the maidenhair tree from China (*Ginkgo biloba*), Japanese maples, Australian willow (*Geijera parviflora*), deodar cedars from the Himalayas, and towering English elms.

This splendid forest served as a background for Phelan's primary interests at Villa Montalvo, namely, entertaining and the arts. The list of guests who came here is dazzling. Visitors, some frequent, included a number of writers, such as poet George Sterling, novelists Gertrude Atherton and Kathleen Norris, Jack London, and short story writer Charles Caldwell Dobie.

Franklin Delano Roosevelt came here when he was assistant secretary of the Navy. So did opera star Lawrence Tibbett, Grand Duchess Marie of Russia, Indian chief Snow-on-the-Mountain, and tennis champion Helen Wills Moody.

Senator Phelan died in 1930, but his will ensured that many of his traditions would live on at Villa Montalvo. He willed the one-hundred-seventy-five-acre property to the San Francisco Art Association, which eventually gave over ownership to the Montalvo Association, a nonprofit foundation. Today the Santa Clara County Parks Department maintains the gardens and the surrounding wilderness area, which is a wildlife sanctuary.

The house and its immediate grounds are run by the Montalvo Association, in keeping with Phelan's wish that Villa Montalvo be used "for the development of art, literature, music and architecture by promising students."

To this end, artists in residence live in the guest apartments near the house, staying two months at a time in this peaceful place. The association also sponsors a writing competition for high school students, a biennial poetry contest, and writers' workshops. A small gallery on the property houses temporary art exhibits.

Most people get to know Villa Montalvo and its stately gardens through the busy summer program of musical and dance events, both classical and pop. On summer evenings, hundreds of music lovers throng on the lawn during the intermissions, a sight that certainly would have pleased James Duval Phelan.

VILLA MONTALVO

Santa Clara County Parks and Recreation Department, 15400 Montalvo Road Saratoga, California 95070 408/867-0190. Concert, Villa, and art show information 408/741-3421. Weekdays, 8 to 5. Weekends and holidays, 9 to 5. Shows in the gallery, Thursdays and Fridays, 1 to 4; Saturdays and Sundays, 11 to 4. Since grounds may be closed during special shows, call ahead. From Highway 17 at Los Gatos, take the Highway 9 (Saratoga Avenue) exit, toward Saratoga. After about 3 ½ miles, turn left on Montalvo Road.

Lush lawns, and towering trees are the framework for Senator James Duval Phelan's classical country house— the Villa Montalvo—making the grounds one of the Bay Area's best-loved settings for summertime musical and operatic programs.

HAKONE GARDENS AND CULTURAL CENTER

Saratoga, California

In 1915, Isabel Stine joined the crowds that flocked to the Panama-Pacific International Exposition in San Francisco. She was so intrigued with the Japanese garden that she went to Japan the following year to spend six months studying its culture. In 1917 Mrs. Stine brought an architect, Tsunematsu Shintani, and an imperial gardener, Naoharu Aihara, to design a Japanese garden on sixteen acres of wooded hillside near the little town of Saratoga.

The result of their collaboration is Hakone, named after a mountain spa on the island of Honshu and considered by some experts the finest hillside strolling garden in the United States.

Included in the original design was a teahouse, constructed without nails in classical Japanese manner, to provide a place to look down upon the garden and to view the summer moon. Five years later she ordered the construction of a second building, the Lower House, with three bedrooms so that she, her husband, Oliver, and their three children could use the property as a summer retreat. Although the house had some Western amenities, the family lived in Japanese style, sleeping on futons unfolded on tatami mats.

Major Charles Lee Tilden bought the property in 1930 and continued to make improvements, including the impressive carved wooden gate to mark the main entrance. In 1960, a group of six families purchased the property, keeping it for six years before announcing plans to sell it to developers. Within twenty-four hours, the Saratoga City Council decided that it was out of the question to let the gardens disappear and promptly purchased the sixteen-acre parcel for $145,000.

Since then, the city and the Hakone Foundation, a citizens' support group, have been conscientious stewards, interested in authenticity and tradition. When they took over, the gardens were overgrown and in poor condition, after several years of negligence. Japanese garden specialist Tanso Ishihara was hired to restore the plantings according to the original design.

The heart of that design is the hillside garden, sloping upward from the entrance towards the Moon Viewing House. Paths lead gradually up the hill, guiding the visitor around the pond and to viewing sites like the wooden pavilion that is curtained with clouds of blue and white wisteria in spring. This is one of the best places to look across the pond and its resident population of koi towards the three-tiered waterfall. In keeping with Japanese tradition, the view of the cascades is partially interrupted by carefully placed trees.

Even when nothing is in flower or the day is dull and overcast, Hakone is interesting to look at, thanks to its varied shapes, foliage colors, and textures. Foliage colors range from the almost black growth of cypress to the tender green of golden bamboo. Textures are equally important, to be appreciated by touching as well as looking. As explained by a guide, the Japanese differentiate between "male" and "female" pines according to texture. The Japanese red pine, with its soft and pliant needles, is considered feminine, while the spiky needles of the black pine are masculine.

In addition to the large hillside garden, Hakone has several specialty gardens, some added since the city began to manage the estate. A small Zen garden, enclosed by a low stucco wall, is next to the Lower House. This is an example of the type of garden created not to be walked in but to be looked at as an encouragement to meditation. A few large rocks, an expanse of raked gravel, and several carefully shaped trees and shrubs are all that it takes to create a landscape that is as much about emptiness as content.

Around the corner, another special-purpose garden serves as an entrance to the tea ceremony rooms inside the house. A curve of stone steps leads through a landscape carpeted with moss and outlined by Japanese maples. At the bottom, participants remove their shoes and wait for the signal to duck under the low door that leads into a room carpeted with tatami mats.

The Bamboo Park was launched in 1987 by the Japan Bamboo Society in cooperation with Saratoga's sister city, Muko-shi, a suburb of Kyoto. The garden features about thirty types of bamboo in a surprising range of shapes and sizes, from bear bamboo, a ground cover from the Japanese Alps, to giant timber bamboo, which grows to more than a hundred feet.

In the midst of the bamboo, a symbolic garden honors the sister city relationship. White gravel represents the Pacific Ocean, with a peninsula of rocks and a lighthouse lantern dramatizing the two countries reaching out towards each other. To one side of this metaphorical ocean are twenty-six stones representing the city council of Muko-shi, while five stones on the other side symbolize the council members of Saratoga.

Traditional gestures like these are typical of the lively program of activities at Hakone, which is now an active cultural center as well as a garden. Its newest building is a re-creation of a nineteenth-century Kyoto merchant's house. The building provides space for parties, ceremonies, and classes. Two traditional Japanese apartments will house artists in residence.

The chosen artists will spend several months working and teaching at Hakone. Like the garden itself, they will be lively ambassadors of Japanese culture.

❦ HAKONE GARDENS AND CULTURAL CENTER
21000 Big Basin Way Saratoga, CA 95070 408/867-3438. Monday through Friday, 10 to 5; Saturday and Sunday, 11 to 5. Closed legal holidays. Donation requested. Parking fee, $3. Take Highway 17/880 or 280 and exit to Saratoga. From downtown Saratoga take Highway 9, which is Big Basin Way.

Bamboo structures
and stone lanterns
are important features
in the design
of the Hakone gardens,
which were originally inspired
by the 1915
Panama-Pacific
International Exposition
in San Francisco.

A teahouse
and a summer cottage,
in traditional style,
are highlights of one
of California's
most elaborate and authentic
Japanese gardens.
Hakone is the setting
for numerous
cultural and educational activities.

ELIZABETH F. GAMBLE GARDEN CENTER

Palo Alto, California

In 1901, a young man named James Gamble and his parents came from Kentucky to look at Stanford University, then a fledgling institution only fifteen years old. The family liked California so well that in 1902, when James returned to Palo Alto to enter college, his father, mother, and three siblings all came along. What this did to young James' plans for independence is unknown, but the legacy of the Gambles' move west is one of the most appealing gardens in Northern California.

Although James' father, Edwin, was a grandson of one of the founders of Proctor and Gamble, he chose not to build an ostentatious estate. Instead the family built a comfortable family house, Colonial/Georgian Revival in style, surrounded by gardens where the children rode their ponies and Mrs. Gamble tended a cow and a flock of chickens. It was a rural neighborhood then, only the second residence south of Embarcadero Road.

Elizabeth Gamble, the youngest of the four children, lived in the house until her death in 1981. She continued to add to the charm of the garden, adding beds and building a teahouse for outdoor receptions. A quiet and generous philanthropist, she willed the house and garden to the city of Palo Alto, which leased it to the community horticultural foundation that now operates the property. The property is divided into formal and demonstration gardens, separated by a small area of woodland featuring shade plants like the oakleaf hydrangea. Working from the original designs, the resident horticulturist, Scott

Loosley, and a host of volunteers have carefully restored the formal garden. It stands as a fine example of Edwardian garden design, a transition between the formality of Victorian layouts and naturalism such as that of English garden designer William Robinson.

The basic design of the garden was created by San Franciscan Walter Hoff, who divided the two-and-a-third-acre lot into highly differentiated spaces separated by hedges and walls. Within these spaces, the planting is soft and romantic in feeling.

The rectangular wisteria garden at the back of the house has become a popular setting for springtime weddings. The small lawn is outlined by wisteria vines trained into low tree shapes. The wisteria is a rarely seen double-bloom variety, so its blooming season, usually in March, is particularly spectacular. More spring color comes from the beds planted with red and white and blue pansies. In summer, when the wisteria trees are covered in green leaves, impatiens, columbines, and foxgloves provide color.

Nearby is the rose garden. Although its space is not symmetrical, the layout of circles within circles gives it a calm harmony. In the center is a grassy lawn, surrounded by a circular bed filled with white roses, including Iceberg and Joey. The outer beds, separated by a pathway, are planted primarily in pastel colors. The specialty here is English roses, which combine the shapes and scents of old-fashioned roses with the repeat blooming of modern hybrids. Roses have a long season in Palo Alto's sunny climate, so that even in late October one can smell the intense fragrance of Mary Rose and Compassion and see buds on Apricot Nectar.

The most ambitious and formal section of the garden is the allée of weeping cherry trees. In spring, one is drawn along the path past solid masses of pale pink blossoms towards a giant palm tree. A bench has been placed at its base to encourage people to enjoy one of the garden's most beautiful views. To the left one sees the house and the front yard, and ahead, the allée of cherry trees leads the eye back towards a fountain and grotto, popular items in early-twentieth-century gardens.

The survival of this period garden is now assured by the Elizabeth F. Gamble Garden Center. Spokeswoman Patricia Polhemus says, "Our role is to provide a beautiful garden. The most important thing is simply that it exists, that in a city that's ambitious and busy, there are two-and-a-third acres of beauty."

In addition, the center wants to teach people about horticulture. To this end, the center's demonstration gardens are a collage of interesting projects, some designed for people who may never have had a chance to grow anything before. The largest is the vegetable garden called Roots 'n' Shoots, a program in which small groups of third-graders and senior citizens work together on a plot for an entire growing season. The project has been so successful that the Gamble staff has put together a curriculum guide for other groups interested in creating similar programs.

The University of California Cooperative Extension puts on regular exhibits, ranging in recent years from a mouthwatering array of salad greens and edible flowers to a down-to-basics lineup of eight different kinds of mulch. An herbaceous border, also in the demonstration gardens, shows off some fifty different plants, including some sun-loving ground covers, like a gray-leafed *Tanacetum haradjanii* from Syria and the furry-leafed, gray-green *Stachys byzantina* from Turkey, popularly and accurately known as lamb's ears.

The Gamble Garden is a lovely and useful enhancement to the city of Palo Alto. Future plans include refurbishing the interior of the house to give visitors an even better idea of what gracious living was like in the early part of this century.

❦ ELIZABETH F. GAMBLE GARDEN CENTER
1431 Waverley Street Palo Alto, CA 94301 415/329-1356. Daily, 8 to dusk. Free. From Highway 101, take the Embarcadero Road exit toward Palo Alto; turn left on Waverley. From Highway 280, take the Page Mill Road exit; Page Mill becomes Oregon Avenue, from which turn left on Waverley. Visit mid-March through mid-April for weeping cherry trees, bulbs, and wisteria in bloom, and in April and May for the old-fashioned roses.

The scarecrow (above)
was made by third graders
and senior citizens
to protect the vegetable garden.

Roses and lavender
surround the circle of grass (top left)
used by the Gamble family
and their friends
for clock golf, a lawn game
resembling croquet,
in the early decades
of the twentieth century.

The Gamble house (bottom left)
is rimmed with magnolias, azaleas,
and camellias.

The Gamble garden
has a gentle,
old-fashioned quality,
epitomized by this unstudied
assemblage of summer flowers
and a simple
wooden fence.
The wisteria garden
nearby is a popular setting
for weddings.

SUNSET GARDENS

Menlo Park, California

Since the early 1950s, *Sunset Magazine's* gardens in Menlo Park have been an inspiration to California gardeners. The velvety acre of lawn and the curving borders filled with flowering plants were laid out by California's leading landscape designer, Thomas Church, in perfect harmony with architect Cliff May's ranch-style office buildings.

From the two-story entrance lobby of Sunset's offices, visitors see an idealized suburban landscape. The huge lawn, shimmering with blue-green highlights in the sun, seems more luxurious than ever after years of drought. The border plantings form a ribbon of color along the edge, with a backdrop of shrubs and trees giving a sense of enclosure. The contours of the garden were inspired by the winding path of San Francisquito Creek.

Thomas Church laid out the border plantings to illustrate the plants of the Pacific Coast from the Mexican border to Canada. Thanks to Menlo Park's kindly Zone 15 climate, Sunset's garden manages to incorporate most of the twenty-four climate zones described in the *Sunset Western Garden Book*, best-selling bible of gardeners from the Pacific to the Rockies. Resident gardeners say that they do push the limits, pointing at a jacaranda tree, a native of Brazil, that suffered considerable damage during the long freeze that began at Christmas in 1990.

Moving clockwise from the main lobby past a long bed filled with jasmine, visitors start the tour of the Pacific Coast at the southern border of the state. A matilija poppy *(Romneya coulteri)*, with its plate-sized white blooms textured like crumpled paper, is typical of California natives that survive even with poor soil and little water.

Nearby, a small area is devoted to desert plants, including small cacti and a variety of succulents. Although most of the plants in this section are modest in size, a rare *Yucca filifera* sends its flower clusters forty feet high.

As one moves along the path, metaphorically ascending the coast, not only the flowers change but the trees and shrubs as well, so that the background planting moves from manzanita and ceanothus to camellias and rhododendrons, and finally to dog-woods, which do best in the cool, damp gardens of the North-west. Note the tall Port Orford cedar, which marks the border between California and Oregon. A redwood grove pays tribute to the Lane family, former ownwers of Sunset; each of the eleven trees represents a family member who was active in the business.

Since the Lane family sold their interests in the magazine, the philosophy of the garden has changed somewhat. In the past, a long-time employee explained, the garden was considered a showcase, the plantings remained the same from year to year, and no expense was spared to make sure that every leaf and blossom was perfect.

Under new ownership, Sunset's grounds have become more of a test garden, with increased synergy between the plantings and stories in the magazine. In recent years, many of these have reflected the readers' concerns about saving money during a recession. In practical terms, this has meant that Sunset's gardeners have been growing more plants from seed, using mulch made from their own downed oak trees, and doing more dividing and propagation in their own greenhouses.

The wall has come down, literally, between the main garden surrounding the lawn and the editorial test garden. This area isn't beautiful, but it is fascinating, offering visitors a chance to see some of the horticultural experiments that may become

stories in the magazine. On a typical day, visitors might find twenty-five pots of different mints or a bed with a couple of dozen different kinds of eggplants. There might be a twenty-five-foot-square garden designed for novice gardeners (tended by an office worker, not a professional horticulturist), some new variations on container gardens, or an array of unusual vines being tested as trellis plants. About one-third of these work out well enough to be featured in the magazine.

Increased concerns about drought and pesticides have also influenced the choice of plants in recent years. The editorial test garden is strictly organic, and pesticide use has been substantially reduced in the main garden. Some lawns have been replaced by drought-tolerant ground covers and shrubs, while some formal plantings—primulas, tulips, and begonias—under the oak trees have been replaced by woodland plants such as campanula, trailing geraniums, ferns, bergenia, and Pacific Coast iris. *Sunset*, and its readers, are taking a greater interest in hardy, drought-tolerant native plants.

Tour guides also take visitors to the enclosed patio, shaded by a jacaranda tree and filled with the container plants that the magazine has made so popular. Cactus, aloes, and dwarf citrus—notably a fence covered with espaliered lemon trees—are also featured in this warm spot.

For all its well-manicured perfection, this is a realistic garden, designed to inspire gardeners rather than to intimidate them. The emphasis is on plants that are generally available to the magazine's readers, and the gardeners make a conscious effort to plant something for everyone so that both novice gardeners and experts will always find something of interest. Judging by the number of visitors, both individuals who drop in and large bus tours, they are succeeding admirably.

❧ SUNSET GARDENS
80 Willow Road Menlo Park, CA 94025 415/321-3600.Weekdays, 9 to 4:30. Guided tours of grounds plus test kitchens and offices, 10:30 and 2:30. For groups of 8 or more, call to arrange a tour: 415/324-5479. Free. From Highway 101 at Menlo Park, take the Willow Road exit west.

The beds at Sunset Gardens
are laid out
to give visitors a walking tour
of different
Pacific Coast climates.
The gardens include
both woodland settings
and a miniature
desert filled
with agaves and cacti.

Sunset Magazine
has always been
sympathetic to the needs
of plant-lovers who
must plant their gardens in pots.
The courtyard
of the magazine's
Menlo Park headquarters
shows off
a variety of container plants.

If a garden is man's attempt to create a world of his own, rarely has he been more successful than at Filoli, one of America's grandest and most beautiful estates. Its setting is the foothills thirty miles south of San Francisco, a wilderness of madrone, redwood, and poison oak, where coyotes roam and rattlesnakes slither through the heavy underbrush. Inside Filoli's walls, all is formality. Lush lawns and perfectly disciplined plants form a series of spaces as precisely delineated as the rooms of a house.

Filoli is a paradigm of the grand estates of late-nineteenth- and early-twentieth-century America, when land was cheap and the very rich could afford as many as thirty full-time gardeners to maintain their elaborate fantasies. East or West, the inspiration for these great estate gardens was always European. For aspiring American aristocrats, recreating European designs gave a comforting sense of tradition.

In 1915, when William Bowers Bourn II began to plan Filoli, California regional architecture had not developed its ideas of informality, native plant materials, and a close relationship between house and garden. Nor did completely naturalistic English gardens hold much appeal; in California, a garden's greatest appeal was its ability to keep the rugged landscape at bay.

Bourn, whose money came from gold mines and from his ownership of the Spring Mountain Water Company, wanted formality in both his house and his garden. He turned to San Francisco architect Willis Polk for the thirty-six-thousand-square-foot mansion in modified Georgian style. For the garden design he chose Bruce Porter, who favored formal Mediterranean style. Each did his work, not so much as a collaboration but as a parallel effort.

As in a formal house, the garden is a series of rooms, each with its own theme; the spaces are separated by brick walls, boxwood hedges, or allées of Irish yew trees. Although Bourn saved the California live oaks that were already on the site, few other plants are native. When he traveled, he brought plants home from around the world; the towering yew trees came from cuttings from his estate in Ireland.

After the deaths of Mr. and Mrs. Bourn, Mr. and Mrs. William Roth purchased Filoli. Lurline Matson Roth, a member of the Matson shipping family, was an enthusiastic gardener who retained the original design of the garden while adding hundreds of camellias, magnolias, and rhododendrons.

Today, some three thousand species of plants furnish Filoli's gardens. As early as February, visitors enjoy the sight of thousands of daffodils and camellias in bloom, the sweet smell of daphne, and the spectacle of some rare magnolias, including the yellow-flowering magnolia Elizabeth, originally developed at the Brooklyn Botanic Gardens, and the *Magnolia campbellii*, whose plate-sized, pale pink blossoms resemble birds in flight. Lilacs, dogwood, iris, and roses come into bloom later in the spring and early summer.

The Sunken Garden is the garden's largest and most formal section, a spacious reception room furnished with closely clipped hedges and shrubs, formally pruned trees, and a carpet of velvety grass, truly a luxury in drought-prone California. Touches of color liven the palette of greens from deep evergreen to silvery gray. A rotating display of annuals blooms in neat beds and borders and in large terra-cotta pots. The brilliant yellow of a sunburst honey locust tree draws the eye down the rectangular pool to the far wall of olive trees and yews.

Each of the smaller rooms of the garden has its own character. The High Place has a distinctly Italianate feeling, with its wall of dark cypress trees and curving colonnade around a small lawn. The columns, it is said, were ballast in the ships that sailed around Cape Horn to San Francisco during the Gold Rush.

The Walled Garden, on the gentle slope just outside the Teahouse, is filled with color provided by masses of annuals planted in large beds and borders. The Knot Garden pays tribute to classical French garden design; here, hedges of santolina, dwarf lavender, and hyssop form a knot pattern. Nearby, the Chartres Cathedral Garden takes its inspiration from a stained glass window, recreating the luminous effect with beds of petunias and roses outlined by boxwood hedges.

Elsewhere, Filoli's collection of 383 different roses includes a wide variety of old-fashioned, climbing, and tea roses. Rose fanciers will appreciate the careful, legible labeling.

Even the most formal garden has its private side, and at Filoli it is the cutting garden, off limits to visitors. The results, however, can be seen in the magnificent flower arrangements in the house, which has been refurnished to give visitors an idea of the grand life. Today, the arrangements are done by some of the eight hundred volunteers who have helped to staff Filoli since Mrs. Roth moved out and donated the central part of the property to the National Trust for Historical Preservation.

Docent tours, flower-arranging demonstrations, lectures on art and architecture, classes in gardening techniques, musical performances, benefit parties, and an attractive shop selling garden-related items make Filoli a vital place in a world that has changed radically since Mr. Bourn built his garden.

❦ FILOLI

Cañada Road Woodside, CA 94062 415/364-2880. Tour the gardens without a guide on Fridays and the first Saturday and second Sunday of the month. On Tuesdays, Wednesdays, Thursdays, and other Saturdays, tours are by reservation. Monday through Saturday at 10, guided nature hikes, also by reservation. Guided or unguided access, $8 per person. No children under 12 admitted. From Highway 280 take the Edgewood Road exit west to Cañada Road; it's about a mile to the gatehouse. The garden is closed to the public from mid-November to mid-February.

Summer is a wonderful time
to admire Filoli's annual
and perennial borders,
delphiniums, and leptospermum.
In the fall, Japanese maples
put on a spectacular
foliage display,
while roses continue
to bloom into November.
Filoli is a paradigm
of the grand estates
of late-nineteenth-and
early-twentieth-century America,
when land was cheap,
and the very rich could afford
as many as thirty
full-time gardeners
to maintain
their elaborate fantasies.

Filoli has been
fortunate in many ways.
It has been
meticulously maintained
since the beginning,
without transition periods
that might have blurred
its demandingly
precise design.
The estate is protected
by hundreds of acres
of watershed land.

JAPANESE TEA GARDEN AT CENTRAL PARK

San Mateo, California

In 1966, when the Japanese Tea Garden at Central Park opened, the town of San Mateo was a sleepy, mostly low-rise suburb. Twenty-five years later, the garden's silver anniversary was celebrated amid the screeching and pounding of construction crews building a solid wall of high-rise office and apartment buildings around the park. This is a small gem of a garden, made even more precious by its location in a burgeoning downtown area.

In spite of the changing urban scene, the garden remains a remarkably serene place, conveying a feeling of safety and self-sufficiency. Its one-and-a-third acres are completely surrounded by a wooden fence. A protective ring of towering live oaks stands sentinel and gives the impression of separating the Japanese garden from the rest of Central Park. Visitors instantly have a sense of being far away from the bustle of shops and offices. Curator Sam Fukudome describes his tiny fiefdom affectionately, saying, "It's not too big, not too small, just right for a little walk."

Like many Japanese gardens in California, this one is the result of a cooperative effort involving the municipality, the local Japanese community, garden clubs, and a sister city in Japan, in this case, Toyonaka. The designer was Nagao Sakurai, former landscape architect at Tokyo's Imperial Palace.

On a small, flat site without even a semblance of a hill, Sakurai managed to create an enchanting landscape, full of artful effects that both echo nature and improve upon it.

Sakurai made the garden
seem much larger than it is
by creating an abundance
of small vignettes.
Visitors should not only
proceed slowly but stop often,
to look behind them
at a landscape that seems
to recreate itself endlessly
depending on one's viewpoint.

The visitor's first view from the gate is a dramatic one. One looks straight ahead across a pond to a five-tiered concrete pagoda, a gift from the city of Toyonaka. Before focusing on this structure, the eye is caught by a colorful combination of trees. In spring, they combine into a palette ranging from the brilliance of azalea blooms to the bright burgundy leaves of Japanese maples to the deep, rusty purple leaves of a flowering purpleleaf plum. From other angles, these same trees fulfill an important Japanese garden convention by forming a frame or partial screen around the waterfall that tumbles down a small hillside near the pagoda.

Sakurai made the garden seem much larger than it is by creating an abundance of small vignettes. Visitors should not only proceed slowly but stop often, to look behind them at a landscape that seems to re-create itself endlessly depending on one's viewpoint. The details are highly refined. Here, a weeping flowering cherry falls with especial grace towards the water; there, a little mound combines the shiny green leaves of a magnolia, the deep red of a maple, and a massive stone boulder with vivid effect.

Home gardeners ambitious to create a similar garden may be both impressed and daunted by the map that locates and lists the garden's trees, because seventy-one different varieties of trees play roles in this small landscape. These include many Japanese classics, such as black and red pines, maples, and flowering fruit trees. The collection of maples features delicate, finely cut varieties such as the bushy *Acer palmatum* 'Dissectum' near the central bridge. The wide selection of maples is one reason that this garden is as spectacular in autumn as it is in springtime.

Other trees on the list that are not normally associated with Japanese gardens were already thriving at the site when the garden was founded. Just behind the verdant display of ferns and azaleas are a couple of Portuguese laurels that have been transformed by being pruned into traditional shapes.

Skillful, constant pruning is the key. Some other Japanese gardens in the Bay Area have lost their distinctive character because of negligence, often spurred by budget cuts, or simple ignorance of the details that create authenticity. Fukudome, a bonsai specialist, points at a large Japanese maple that he pruned a month before, saying that he must prune it again within a month. After the spring blossoming of the azaleas, he picks off thousands of dead blooms by hand. When he clips the thread cypress, people gather to watch his precise cuts and to ask advice.

Visitors to the garden should also note the planting outside the front gate, in front of the park's recreation center. The beds, designed by Sadao Sugimoto, feature some outstanding specimens. The goldenrain tree (*Koelreuteria paniculata*) lives up to its name when it produces masses of bright yellow flower clusters in spring and summer. Low clumps of heavenly bamboo (*Nandina domestica*, not a bamboo at all) have variegated leaves in pale green, wine, and bronze tones. A tree wisteria is covered with massive white blossoms in April, welcoming passersby to the garden.

JAPANESE TEA GARDEN AT CENTRAL PARK
Mail: 330 W. 20th Avenue. Location: 50 East Fifth Avenue San Mateo, CA 94403
415/377-3345. Monday through Friday, 10 to 4. Weekends and holidays, 11 to 4. Free.
From Highway 101 at San Mateo, take the Third Avenue exit west, to El Camino Real;
turn left, and left again on Fifth Avenue. Visit in April for the cherry blossoms and in
October for fall foliage.

Ever colorful,
trees in spring range
from the brilliance
of azalea blooms
to the burgundy leaves
of Japanese maples.

49

STRYBING ARBORETUM AND BOTANICAL GARDENS

San Francisco, California

Strybing Arboretum and Botanical Gardens' seventy acres along the south edge of Golden Gate Park enclose many smaller gardens, groves, and meadows, each dedicated to the plants and trees of a different part of the world. The mission of each section is to show home gardeners the wide range of species that will survive and thrive in the Bay Area's climate, sometimes described as Mediterranean-plus-fog.

"This is a municipal garden," points out Walden Valen, Strybing's director, "and our purpose is to serve the public that pays the taxes." First and foremost, Strybing is one of the loveliest places in Golden Gate Park, a place for children to see swans adrift on a pond, for older people to sit surrounded by flowers and sunshine, for walkers to take a tour of the world—or at least the portion of it between forty degrees latitude north and south—on foot.

Starting at the main entrance and proceeding counter-clockwise, a visitor's first stop is the demonstration gardens just to the right of the front gate. Here, the trees include a Japanese cherry tree (*Prunus serrulata*), the Australian peppermint tree, and the reddish coral-bark maple (*Acer palmatum* 'Sango Kaku'). The beds are changing displays of Strybing's most successful plant testings, planted en masse to show potential customers exactly what their seedlings will become.

The Jean Wolfe garden, a tribute to a local designer who was a protégée of Thomas Church, shows what can be done on a small, irregularly shaped lot without spending a great deal of money. It is a charming green-and-white garden with an eight-sided pavilion designed by Church.

Two of the garden's most popular attractions, are the Garden of Fragrance, whose low-walled design was created to make it accessible to the blind, and the Biblical Garden, planted with species mentioned in the Bible or native to the eastern Mediterranean.

Continuing along the path, the visitor passes a plentiful array of azaleas to reach two of Strybing's major collections, magnolias and rhododendrons. Strybing specializes in vireya and Madden rhododendrons. The Vireyas, native to the mountains of Southeast Asia, come into bloom throughout the year, but the fragrant Maddens are at their best in April.

Next on this walk are several gardens representing various parts of the world with Mediterranean or temperate climates. These include the Cape Province of South Africa, an area with a Mediterranean climate and a rich diversity of plant life. Eye-catchers include mounds of pincushion flower (*Leucospermum cordifolium*), proteas, and agapanthus, a handsome blue-flowering plant so well adapted to California that it is planted in the center divider of many freeways.

In the next few yards, the visitor leaps across oceans to arrive in New Zealand, represented by the massive New Zealand Christmas tree (*Metrosideros excelsus*) and the droopy rimu tree (*Dacrydium cupressinum*), a conifer with berrylike fruits instead of cones. Just a bit further is Eastern Australia, a collection of drought-resistant plants that seem particularly appropriate to California these days.

Strybing is continually trying to expand gardeners' ideas of what will grow in the Bay Area. Two of the newest areas are Chile, whose Mediterranean climate resembles that of coastal California, and the New World Cloud Forest, sheltering some of the orchids and bromeliads of high-altitude tropical areas. Both are in the development process, and the cloud forest suffered considerable damage in the unusual spell of freezing weather that occurred in December 1990 and January 1991.

Horticultural snobs who concentrate on rare and exotic plants may change their minds when they see Strybing's meadow devoted to California natives. Landscape architect Ron Lutsko Jr.'s award-winning design combines swaths of wildflowers with the bunch grasses that used to cover the state before cattle and European grasses came along. In spring, the meadow is a tapestry of color, provided by pale lavender iris, orange poppies, deep blue ceanothus, and pale blue *Nemophila menziesii* (baby blue eyes). One stands in the middle and hopes that California really looked like this a century ago.

Strybing puts considerable emphasis on design. The gardeners are skilled at combining different plants, so that the public gets ideas on how to use a given species attractively. In the interest of home gardeners, Strybing is willing to plant cultivars, unlike other botanic gardens that concentrate on documented specimens gathered in the wild.

Strybing's user-friendly attitude is also expressed in its clear labeling, monthly sales of hard-to-find plants, tours led by knowledgeable docents, seasonal self-guided tour leaflets, a bookstore carrying current gardening titles, and the Helen Crocker Russell Library, an excellent facility for research on all things horticultural. Even visitors who barely know a tulip from a rhododendron cannot help but learn and enjoy themselves here.

STRYBING ARBORETUM AND BOTANICAL GARDENS
Golden Gate Park Ninth Avenue at Lincoln Way San Francisco, CA 94122
415/661-1316. Weekdays, 8 to 4:30; weekends and holidays, 10 to 5. Free. The arboretum is just inside Golden Gate Park at the Ninth Avenue and Lincoln Way entrance, near the Music Concourse.

Several gardens
represent various parts
of the world
with temperate climates.
These include
the "Cape Province
of South Africa,"
an area in the garden
with a Mediterranean climate
and a rich diversity
of plant life.

JAPANESE TEA GARDEN

San Francisco, California

For almost a century, San Francisco's children have been darting along the convoluted paths of the Japanese Tea Garden and shrieking with delighted fear as they descend the steep drum bridge. Adults, too, sense that it is a magical kingdom, a miniature landscape filled with surprises, both man-made and natural.

Makoto Hagiwara, the ingenious landscape architect who created the gardens for the Exposition of 1894, wanted to convey the feeling of a Japanese mountain village. Instead of the carefully conceived vistas characteristic of more formal Japanese gardens, he created a jumble of small-scale mountains, forests, and lakes.

Here the walls are often thick stands of bamboo, sometimes allowed to grow in towering masses, in other places cut short into artful curves that add interest to the beds of flowering shrubs. Steep flights of stairs and rounded mounds covered with shrubbery reinforce the feeling of being in a mountain landscape.

If this fantasy village has a downtown, it is the confluence of paths just inside the entrance gate. Visitors funnel in and out, stopping to decide which path to follow or to take a final photograph. The busy teahouse is directly in front of the gate. While this little building is hardly a serene setting for a ritual tea ceremony, it's a pretty place to sit.

The little pond below the teahouse was designed by Nagao Sakurai, a Tokyo University professor who designed several Japanese gardens in California, including the one in Central Park in San Mateo. In spring this tiny vista is an introduction to the garden's predominantly pink color scheme, expressed in both flowers and foliage. Around Sakurai's tiny lake, the pale pink of cherry blossoms shades into the bright hues of azaleas and camellias and the muted magenta of maple leaves.

To the left, as one enters the main gate, is the striking drum bridge, a scaled-down version of the steep, semicircular bridges that allow the passage of high-prowed pleasure boats along Japanese rivers. Just beyond is the garden's main pond, an undulating expanse of water populated by real ducks, real koi fish, and a pair of startlingly realistic bronze cranes.

A steep flight of steps leads upward from the pond, passing a collection of dwarf conifers. At the top of the hill an open plaza is dominated by a ceremonial gate and a five-tiered pagoda painted in brilliant shades of red and orange. Both were originally constructed for the Panama-Pacific International Exposition of 1915 and moved here in 1916.

The quietest part of the garden lies behind the pagoda. Even on a sunny spring afternoon, few visitors take the cobblestoned path that leads through walls of bamboo and camellias to a small Zen garden. In the traditional style designed to encourage meditation, the flow of gravel is a metaphor for water, with rocks and plants placed to represent islands and mountains.

The Japanese Tea Garden is an integral part of San Francisco's history, not only in terms of its longevity and popularity, but as a microcosm of the city's complicated relationships with its Japanese population. The idea of including a Japanese garden in the 1894 exposition apparently came from George T. Marsh, a dealer in fine-quality Oriental goods and antiques. The garden itself was created by a local landscape gardener, Makoto Hagiwara, a descendant of Japanese nobility. Hagiwara had been trained in the art of landscape design since his youth and had considerable financial resources, thanks to his family's ventures

in silk and rice wine. After the exposition, Hagiwara was granted the concession to run the garden.

In spite of his aesthetic and financial contributions, Hagiwara was referred to in some early accounts as "a Japanese laborer." Anti-Japanese sentiment ran so strong that the Hagiwara family lost the concession between 1900 and 1908. Finally, Golden Gate Park superintendent John McLaren went to Makoto Hagiwara to ask him, humbly, to take charge of the tea garden once again.

Hagiwara took back the concession and moved his fifteen-room Japanese-style house into the park. He and his family nurtured the garden and served tea to decades of guests until 1942, when the Hagiwaras were abruptly sent off to a relocation camp during World War II. The garden was renamed the Oriental Tea Garden and the Hagiwaras' house and family shrine were destroyed.

Nevertheless, the Hagiwara family returned to their beloved garden after the war was over. Even after Hagiwara's death, his stepdaughter continued to operate the teahouse, serving tea with great charm until her death in 1972.

Today, the family is commemorated by a plaque near the entrance gate, facing the earlier plaque that gave all credit to George Marsh, and the road outside has been named Hagiwara Tea Garden Drive. In spite of drought and crowds and cutbacks in the park's maintenance budget, the garden remains a place much loved by both tourists and residents.

❦ JAPANESE TEA GARDEN
Mail: Recreation & Parks Department McLaren Lodge Golden Gate Park San Francisco, CA 94117 415/666-7107. From March 1 to September 30, open 9 to 6:30; free from 9 to 9:30 and 5:30 to 6:30. From October 1 to February 29, open 8:30 to 6; free from 8:30 to 9 and 5 to 6. Free also the first Wednesday of the month. When ticket booth is open, admission for adults is $2; for seniors and children 6 to 12, $1. In Golden Gate Park on Hagiwara Tea Garden Drive at South Drive, on the north side of the Music Concourse. Visit in April to see the cherry blossoms.

No obvious route
directs you
through the garden.
Visitors choose
among myriad winding paths
and bridges
without knowing exactly
where they will end up.

The biggest pond
is bordered by a variety
of traditional
Japanese favorites,
including azaleas, ferns,
magnolias, irises,
and flowering fruit trees.
The striking drum bridge
over the pond
is a scaled-down version
of the steep,
semicircular bridges
that allow the passage
of high-prowed
pleasure boats along
Japanese rivers.

MORCOM AMPHITHEATER OF ROSES

Oakland, California

"This is one of the best-kept secrets in Oakland," says rosarian David Skinner, caretaker of the Morcom Amphitheater of Roses. In spite of its location a block away from the shops on Grand Avenue, there's not even a sign to point the way to this lavish display of roses.

Morcom, Oakland's official rose garden, has been here since 1934. Much of the original construction work was done by Works Progress Administration workers. They were blessed with a particularly appealing site, a canyon that shelters both visitors and roses from the wind and the sounds of the city. The classically Italianate design emphasizes the long, narrow shape of the canyon. From the entrance, which is flanked by a pair of curving colonnades, the visitor sees a strongly symmetrical design, with a central axis paralleled by neat beds and walkways.

From late March to November, the strict outlines are softened by the blooms of thousands of rosebushes, creating a tapestry of pinks, reds, and yellows. From the start, Morcom has served as a showcase for All-American Rose selections. These are the varieties that have been tested and found worthy by the American Rose Society. By the time the society announces its choices in the spring of each year, Morcom has usually had the winners in the ground for six months to a year, so that gardeners can go immediately to see the new stars in full bloom.

Although the All-Americans are elected for such worthy qualities as resistance to disease and insects, a pretty face also counts a

Roses are
so sensitive to their microclimate
that the same variety
may do well
in one spot
and do poorly fifty feet away.

lot with the judges. The winners are often the showiest and most exuberant varieties. At Morcom, visitors stop in their tracks at the sight of varieties like Headliner, whose variegated pink-and-white blooms measure five to six inches across, and Perfect Moment, whose red blooms with yellow centers are almost as large.

Although Skinner occasionally has to make room for the newcomers by eliminating some of the less popular or more sickly specimens, he estimates that Morcom has about five hundred All-American Rose selections on display. Some are so old that they are no longer commercially available. In spite of the numbers, Skinner seems to know each rose by name and personality. He points out that roses are so sensitive to their microclimate that the same variety may do well in one spot, poorly fifty feet away.

The triumph of Morcom is not only that all these prima donnas bloom enthusiastically but that they all bloom at once. The target day is Mother's Day in early May. On Mother's Day, a ceremony at the rose garden honors Oakland's Mother of the Year, a woman chosen for her family life and community involvement. A metal plaque bearing her name is set into the central walkway of the garden.

Mother's Day is also the start of the garden's busy schedule of weddings. For the next five months, until October, between one and five wedding ceremonies take place here each weekend.

The customary site for the rites is at the top of a double stairway that rises at a right angle from the central axis of the garden. Masses of pink Pride of Oakland rosebushes flower on either side of a series of terraced pools.

Morcom, like most municipal gardens, has had to contend with painful budget cuts in recent years. In spite of cutbacks that reduced the staff from six gardeners to two, the garden is meticulously maintained. The secret is the help of participants in On Trac, a vocational program for mentally and physically disabled young adults. Skinner speaks of them with pride and affection.

"They come here every day all year long, and it really helps tremendously. The park was completely overgrown before they came." Thanks to On Trac members' weeding and raking, Morcom is one of the best-maintained public parks in the area.

It's also a park with a bright future. Thanks to a municipal bond issue, Morcom is scheduled to be completely refurbished in 1993. The original layout will be retained, since this is a historical landmark, but the design will be much enhanced by new turf, appropriately Italianate benches, and pathways paved in bricks and cobblestones. In addition, plans call for nighttime lighting and a drip irrigation system.

With luck, there will also be a new sign on Grand Avenue, directing garden-lovers to this unheralded floral display.

❧ MORCOM AMPHITHEATER OF ROSES
Mail: 1520 Lakeside Drive Location: 700 Jean Street Oakland, CA 94612 510/238-3187. Daily, dawn to dusk. Free. Morcom will be closed for equipment repair until late 1993. From Highway 580 at Oakland, take the Grand Avenue exit east; from Grand take Jean Street left one block. Bloom is fullest in summer.

The triumph
of Morcom
is not only that
all these prima donnas
bloom enthusiastically
but that they
all bloom at once.

65

the seats in an ancient Greek theater. At the very top, a curving wooden pergola is covered with climbing roses, which are usually in bloom by mid-April, a few weeks before the rest of the garden's flowering in May. Most of the roses that festoon the pergola are in sunset shades of yellow, gold, peach, and pink. Varieties include Royal Sunset, Joseph's Coat, Golden Showers, and High Noon.

Below the pergola, descending towards a small rectangular pool, the uppermost tier is planted in roses with deep red blooms. Each tier below is planted in lighter colors, so that one proceeds from deep reds into bright reds and oranges, then yellows and bicolor varieties, down to the pale pink of Sally Holmes. At the very bottom, where brides often stand for their wedding ceremonies, the blossoms are the pure white of Sweet Afton.

It's not just the spectacular hybrid teas, grandifloras, and floribundas that get respect here. Although the central, tiered section catches the eye with its array of showy blooms, Russ Battle has used the rest of the garden to show off a wide variety of other roses. The collection also includes creepers, climbers, shrubs and hedges, old-fashioned garden roses, hybrid musk roses, and rugosas. Battle refers affectionately to a certain side section as Lower Slobbovia, where he features roses for people who don't have a lot of time to spend in their gardens. These easy-to-grow varieties include Red Balls, Mickey Mouse, and Martin Frobisher, which is not only indestructible but fragrant as well.

In the mid-1980s, the city of Berkeley decided to phase out the use of highly toxic chemicals in the garden. Many of the existing plants did not do well under the new program and had to be replaced or moved. Their replacements were chosen with special attention to hardiness and an ability to flourish without toxic chemical treatments.

Beyond these requirements, Russ Battle has a special fondness for unusual, often old-fashioned roses that are not widely available in nurseries and catalogues. A bed that was once devoted to All-American Rose selections, those roses deemed winners by the American Rose Society, has been gradually replanted in less-commercial varieties.

Increasingly, he opts for fragrant varieties, saying, "As much time as people spend looking at roses, they spend smelling them." He points to Jacques Cartier, Intrigue, and Cherry Vanilla as sweet-smelling examples. Green, glossy leaves are another plus when he makes his selections.

Thanks to the wide variety of plants and the mild climate, something is in bloom here most of the year, particularly from April to November. In spite of ravenous deer, who bound in from nearby Tilden Regional Park to sup on the tender first blooms, and urban aggravations like vandals and plant thieves, the Berkeley Rose Garden bursts into spectacular bloom each spring, a minor miracle much treasured by its neighbors and friends.

BERKELEY ROSE GARDEN
Mail: 2180 Milvia Street, 3rd Floor. Location: 1201 Euclid Avenue Berkeley, CA 94704 510/644-6530. Daily, sunrise to sunset. Free. From Highway 80 at Berkeley, take the University Avenue exit. From University turn left on Martin Luther King Jr. Way, right on Cedar, and left on Euclid. Roses bloom from May through October, most fully from June through September.

Rose fanciers delight
in the many varieties
that thrive in the Berkeley
Rose Garden.
Color, scent, glossy foliage,
and natural resistance
to pests and disease are all factors
in the process
of selecting the plants.

UNIVERSITY OF CALIFORNIA BOTANICAL GARDEN

Berkeley, California

Visiting the University of California's Botanical Garden is like opening the *Encyclopedia Britannica:* You can't deal with either all at once. Like the encyclopedia, the botanical garden is crammed with facts, in the form of some twelve thousand species, almost all collected in the wild.

This is the fifth largest botanical garden in the United States. The university's plant experts have been collecting specimens since 1890, when the garden opened at an on-campus location that is now the site of the Moffitt Library. In the 1920s, the garden was moved—plants and all—to its present thirty-four-acre location almost a mile east of the main part of the campus.

The site in Strawberry Canyon is ideal for growing plants from many different climates and altitudes. Rhododendrons from the Himalayas and the cacti of the Sonoran desert coexist happily, thanks to temperatures ranging from about thirty-five degrees most winters to eighty degrees in the summer, a variety of exposures, summer fog, and an absence of wind.

As befits a garden closely affiliated with a university, UC Botanical Garden's mission is to educate its visitors as well as to give them something pretty to look at. Rare, endangered, strange, and just plain ugly plants are given considerable respect. For example, the plant that greets you in the Desert and Rain Forest greenhouse is not a pretty sight, with its two or three flat, gray-green leaves oozing like toothpaste from a woody center. One becomes more fond of it after reading the detailed signs explaining that the *Welwitschia mirabilis* may grow only two leaves but lives for more than a thousand years, only in the Namib Desert of southwestern Africa.

Also in this same greenhouse is an exhibition explaining convergent evolution, the tendency of unrelated, geographically distant species growing in similar conditions to take on similar characteristics. The display of American and South African desert plants proves the point, comparing the similar forms of American agaves and cacti with South African aloes and euphorbia.

Further on to the left, the South African section brightens the hillside with orange, magenta, yellow, and pink blooms as early as February. Early bloomers include numerous aloes, such as *Aloe abyssinica*, with its brilliant orange flowers, and the magenta *Tulbaghia fragrans*.

Across the pathway is one of the garden's most dramatic sections, the New World Desert. Note the barrel-shaped *Ferocactus histrix*, with its two-inch spines, and the curious *Espostoa nana* from Peru, with a protective coverage of long gray hair. April visitors will be rewarded by the sight of cactus in bloom.

Attention-getters in the adjacent Asian garden include the giant *Paulownia glabrata*, a tree from China, and the *Rhododendron grande*, a native of the high mountains of Sikkim, Bhutan, and India, with enormous white flower clusters. Early springtime also brings the brilliant coral blooms of *Rhododendron scabrifolium*.

Nearby, the New Zealand section features a rare tree fern, *Dicksonia squarrosa*, and a rare, unusually handsome tree, *Agathis australis*. Farther along the pathway are the plants of Chile and Argentina, which share a Mediterranean climate with California and are thus sources of plants that thrive here. The visitor will see a lot of plants here that can't be seen at other Northern California botanical gardens. For instance, there's a section devoted to European-Mediterranean plants.

Even with roses, UC takes an educational approach. Its rose garden, a simple, semicircular area with trellises and a pergola, is designed to illustrate the genealogy of hybrid tea roses. A diagram shows that the outer beds are planted with the European and Asian ancestors of modern roses, while the inner beds illustrate the development of hybrids following the combining of the two strains in the middle of the nineteenth century.

Proceeding clockwise down the hill, one passes an area devoted to North American plants. This section is an opportunity for westerners to see what grows in the rest of the country and to learn about unfamiliar faces like prairie golden-rod (*Solidago*) from Kansas and Joe Pye weed (*Eupatorium hyssopifolium*) from Massachusetts.

After inspecting the medicinal gardens, both European and Asian, and the economic plants, visitors may feel they can't take another step. This is the time to head for the lawn, where students read and sunbathe, families picnic, and couples flirt. The cool grass is surrounded by crabapple trees, rhododendrons from the Himalayas, and the towering *Cryptomeria japonica*, a conifer with new growth in a lovely combination of mauve and coral needles.

Strength revived, the visitor can continue to the large area devoted to native California plants. One of the most interesting sections is devoted to the unusual plants of the Channel Islands. For a change of pace, continue on to the Mather Redwood Grove and Miocene Sequoia Forest across Centennial Drive from the main part of the garden.

The best way to appreciate the UC garden's abundance is to return in different seasons, concentrating on those areas that are at their best. Daily tours by knowledgeable docents lead visitors to the most interesting areas. And don't forget to read the signs; remember, the University of California really wants you to learn while you're here.

🌺 UNIVERSITY OF CALIFORNIA BOTANICAL GARDEN
Centennial Drive, University of California at Berkeley, Berkeley, CA 94720 510/642-3343. Daily except Christmas, 9 to 4:45. Free. From behind Memorial Stadium on the UC Berkeley campus, take Centennial Drive ¾ mile.

One of the garden's most
dramatic sections,
the New World Desert,
is punctuated
by cacti and agaves
of all shapes and sizes.
After all this aridity, visitors will
welcome the adjacent section,
the Asian garden
verdant with rhododendrons
and camellias.

The marble
dedication stone (5' high),
found in the Chinese
Medicinal Herb Garden
within the Asian section,
was a gift
from the Ghangzou College
in China and commemorates
the cooperative effort
between the two institutions
and the American College
of Traditional Chinese Medicine
in San Francisco
(see full view overleaf).

TILDEN PARK BOTANIC GARDEN

Berkeley, California

The state of California stretches hundreds of miles from the lush redwood groves near the Oregon line to the desert along the Mexican border. In between these reaches lie most of the world's climatological extremes, with vegetation to match each one. It would take even the most obsessed botanist decades to see all of California's plant life in the wild. Happily, there's an alternative, the Botanic Garden at Tilden Regional Park.

The garden, located in Wildcat Canyon, specializes in California's native plants, displaying several thousand species. As the brochure boasts, Tilden encompasses one hundred sixty thousand square miles in ten acres.

One might be overwhelmed by the sheer numbers of plants on display if the garden were not so well designed for both education and aesthetics. The sloping space is divided into nine major areas, corresponding to natural areas of the state, interspersed with small subsections that show off plant life in specific environments, such as ponds, dunes, or granite outcroppings.

Each plant has an easy-to-read, informative sign, telling not only the popular and Latin names but also whether a plant is rare or endangered, where this particular specimen was found, and the elevation at which it flourishes. In addition, the signs are in a different color for each section, so that one can tell at a glance which plants grow together naturally. Occasionally, signs in two different colors mark those transitional areas where different types of plants coexist.

California native plants, including the state's official flower, the golden poppy, thrive in the gardens of Tilden Regional Park. The gardens recreate many of the state's climatological and topographical zones, from the high Sierra to the desert.

One of the most dramatic areas is the redwood section. One looks up, of course, to gauge the height of the big trees. In this case, the trees are *Sequoiadendron giganteum*, easier to grow and almost as tall as the more famous coast redwoods (*Sequoia sempervirens*). One also looks down to appreciate handsome smaller plants that thrive in the redwoods' shade. These include trillium, licorice fern, the evergreen huckleberry with its reddish leaves, azaleas, and the handsome red alder tree.

We tend to think of rain forests as being strictly a phenomenon of the tropics, but certain areas of Northern California that receive up to ten feet of rain in normal years also qualify for the title. Here, the Pacific rain forest section, lush with huge ferns, gives a vivid idea of a cold-weather jungle.

Not far away is a starkly contrasting environment featuring the desert-loving plants of the Channel Islands off the Southern California coast. Notable here are several rare and endangered species, such as the Catalina hard-tack (*Cercocarpus traskiae*) and the island mahonia. Note, too, the San Ysidro and prickly pear cacti, the highly ornamental white sage with its long, gray-green leaves, and the twisted branches of the five-foot-tall giant coreopsis.

Tilden specializes in two of California's most widespread plants, the manzanita and the ceanothus, which thrive in several of the subsections. Anyone who has hiked a trail in California has seen manzanita, but it's interesting to see the variety here, including the Pajaro manzanita, with sweet-smelling clusters of tiny pink flowers, and the greenleaf manzanita, with its deep red, twisting branches. Ceanothus, also known as wild lilac, is a hardy and drought-resistant plant, covered in March and April with clusters of flowers in white and many shades of blue.

One of the garden's most interesting specialized collections is the re-creation of a granite outcropping, planted with the tiny flowers that manage to thrive on this inhospitable medium. Note the tiny lace fern (*Cheilanthes gracillima*), the rare Tulare horkelia, and the pretty, early-blooming Sierran firespot

(*Nemophila maculata*), with one deep purple spot on each pale lavender petal.

Another small patch has been designed to show East Bay residents what these hills looked like before the Europeans came. Then, the slopes and valleys were covered with meadow barley, California fescue, yarrow, morning glory, soap plant, and a variety of bunch grasses. Later, these natives were crowded out by such aggressive newcomers as wild oats, Italian rye, thistles, and mustards.

The Sierra meadow, complete with burbling stream and small ponds, offers an opportunity to sit in the middle of the garden and to walk in the shade of quaking aspen, narrow-leaf willows, and western junipers. Here, too, is an example of one of California's most intriguing trees, the Great Basin bristlecone pine from the highest altitudes of the White Mountains. This is known as the oldest plant in the world, with some specimens believed to be five thousand years old.

The joy of Tilden Park Botanic Garden is not only the fact that it's such a successful walk-through botany textbook. It's an introduction to parts of California many of us will never see, and it's impossible not to be awed by the sheer multitude and diversity of plants that call California home.

❦ TILDEN PARK BOTANIC GARDEN
Wildcat Canyon Road and South Park Drive Tilden Regional Park Berkeley, CA 94708 510/841-8732. Daily except Christmas and New Year's, 10 to 5. Free. From Highway 580, take Highway 24 east and, just beyond the Caldecott Tunnel, the Fish Ranch Road exit. Continue, uphill, to Grizzly Peak Boulevard; turn right, and then on South Park Drive right again, to the Wildcat Canyon Road intersection.

Manzanitas, which thrive in much of California, are brought together in great variety at Tilden.

A redwood grove gives visitors an idea of the stately forests along the northern coastline.

ture, began to design the garden.

Mrs. Blake's passion for collecting rare species was tempered by Miss Symmes' design to avoid a hodgepodge of unusual plants. Miss Symmes' firm but gentle hand imposed an orderly design on the rough slope east of the house, creating a formal Italianate garden, fashionable during the 1920's. She also helped organize other parts of the garden to eventually accommodate some 2,500 species of plants, many of which still survive to enrich the experience of today's students and visitors. Two of Mrs. Blake's favorites include the rare *Eucryphia nymansensis,* with its bell-shaped white flowers, and *Ilex paraguayensis,* whose leaves produce the South American tea called maté.

The spring on the upper edge of the property became an integral part of the formal garden, providing water for the reflecting pool. Delicate plants and soft, layered effects rather than masses of single colors contrast the architectural forms of hedges and an allée of southern magnolias flanking the reflecting pool. In spring, today's visitors delight in fragile woodland flowers such as hellebores, campanula, freesia, blue bells, and columbine peeking out from the soft greenery of *Oxalis oregana,* with its shamrock-shaped leaves.

Although the layout is Italianate, certain details reflect the family's fascination with Asia. Two Japanese ceramic statues, one representing the sun, the other the moon, face each other across the lawn; and a ten-foot-high ceramic pagoda stands in a circular reflecting pool between the formal garden and the redwood canyon, which lies to the east of the house.

The sunny steep slope west of the house is organized with stone-walled terraces and diamond-shaped beds above a lawn in a bowl, formerly a quarry. This driest part of the garden features drought-tolerant plants from the Mediterranean, Latin America, and South Africa. Dramatic specimens include pride of Madeira (*Echium fastuosum),* with its tall purple flower stalk towering above a round ball of leaves, and colletia, whose stiff,

thorned, wing-shaped leaves make it one of the world's most effective barrier plants.

Even for people of means like the Blakes, maintaining a large and ambitious garden was not always easy. During the Depression finances were tight. A decade later all the ablebodied gardeners went off to World War II, and the garden became overgrown and ragged.

They came up with a solution that was both generous and ingenious. In 1957 they donated their property to the landscape architecture department of the University of California, reserving the right to live there until their deaths. University funding and supervision by a talented and strong-minded landscape architect and faculty member, Geraldine Knight Scott, redesigned much of the garden to reflect its new, more public, role as an educational resource. Since 1958, many modifications and improvements have been made by the landscape architecture department to demonstrate such important design issues as water conservation and the use of native plants.

Although the original design has been respected, the university has not treated Blake Garden as an artifact to be preserved intact. Some parts of the garden, such as the rose garden, have been replanted, often reflecting a greater concern for water conservation. New areas are being cleared and planted, so the garden is bigger than ever.

Despite drought, two severe freezes in 1972 and 1990, an invasion of deer, and a limited budget, the basic framework of Blake Garden continues to pay tribute to the naturalistic spirit of Mabel Symmes and Anita Blake.

❧ BLAKE GARDEN

70 Rincon Road Kensington, CA 94707 510/524-2449. Monday through Friday except university holidays, 8 to 4:30. Free. From Highway 80, take the Albany exit and head east on Buchanan Street, which leads into Marin Avenue and Arlington Circle. From the left side of the circle take Arlington Avenue 1 ⁸/₁₀ miles to Rincon Road, and turn left. Visit in October for the fall foliage.

A formal reflecting pool (top left),
flanked by evergreen magnolias,
a tiled grotto,
and sweeping views
of San Francisco Bay are among
the elements
that make Blake Garden
one of the most appealing settings
in Northern California.
Anita Blake's style has been
called "feminine"
and even "lovable."
She showed great sensitivity
to the natural features of the site,
incorporating them
into the overall scheme.

Even the most formal
design elements
at Blake Garden,
such as this square reflecting pool,
are softened by delicate,
naturalistic plantings
in gentle color schemes.

Achieving this type of garden takes character. The first requirement is patience, because many of these plants are slow-growers, to say the least. Mrs. Bancroft waited twenty years to see her first giant yucca bloom. At eighty-three she looked wryly at her four-foot-high saguaro cactus, commenting, "They say it takes thirty years for them to branch."

Growing unusual varieties of succulents and cacti also means considerable risk of failure. "Everything I planted I looked up, but these plants don't always follow the rules. I think that I probably have had as many failures as successes."

The gardening process isn't easy. Even the most ardent weeders are wary of plunging their hands into plants that are prickly, hairy, thorny, sharp-edged, and stiff. Many of the garden's specimens are large and heavy. When a large agave dies, it takes two men a couple of days to cut away the leaves and a crane and a truck to remove the rest of the rosette. None of this daunted Ruth Bancroft, a full-time gardener who usually spends all day working outside.

Landscape designer Lester Hawkins had never done a succulent garden (he and his partner, Marshall Olbrich, are best known for their romantic Western Hills garden), so he created the design, provided some of the plants, and then left the planting to the owners.

Most of the plants come from Southern California. Mrs. Bancroft recalls, "My husband and I used to drive down there with the kids on vacation, and we would go to all the nurseries and fill up the car with plants."

As a former architectural student, she chose her specimens with an eye towards color and shape. The color palette is a subtle range of pale greens, grays, and soft blues, flashed occasionally with brightly colored, short-lived flowers. The buildings were painted in a soft green inspired by the leaves of certain agaves.

The dramatic shapes of the largest specimens are the skeleton of the garden's design. Agave plants, some as high as fifteen feet, are placed throughout the garden. Many of these sprawling rosettes are made up of thick, blue-gray leaves, contributing to the overall color scheme of the garden. Others are less subtle; the variegated leaves of *Agave americana* look like ribbon candy in their dramatic stripes of brilliant green and yellow.

More drama comes from the yuccas. Some are compact balls of spiky leaves close to the ground, but the most eye-catching grow into tree forms as high as thirty feet. They are most spectacular in spring, when they blossom forth with flower clusters as long as three and a half feet, covered with creamy, pale yellow blooms.

Tall cacti add verticality to the garden. In addition to her lone Saguaro, Mrs. Bancroft has planted similarly spindly columns from Latin America, particularly Chile. Fat cylinders of golden barrel cactus and the flat Mickey Mouse ears of prickly pears (*Opuntia microdasys*) are other distinctive shapes in this arid landscape.

Thanks to a 1988 visit from New York gardeners Frank and Anne Cabot, the long-term future of the garden now seems assured. Upon seeing the garden, the Cabots realized what a great loss it would be if it were swallowed up by the subdivisions that surround it. They came up with the idea of the Garden Conservancy, a national organization devoted to the preservation of some of this country's great gardens.

The Garden Conservancy now operates the part of Ruth Bancroft's land that is devoted to cacti and succulents, in accordance with a conservation easement. Thanks to the new arrangement, the garden is now open to the public two days a week, and plans for a visitors center and horticultural library have been drawn up. Don't be surprised to see Mrs. Bancroft out among her desert plants: She has no intent to stop tending her garden.

🌿 RUTH BANCROFT GARDEN
Mail: 4104 24th Street, Suite 111 San Francisco, CA 94114. Location: Walnut Creek 415/824-2919 (reservations). Mid-April through October, open for tours on Friday and Saturday at 10 and 1, by reservation. Months and hours of operation will change as this garden "in transition" is developed for public use. Donation of $3 requested.

Ruth Bancroft's house,
like other structures in the garden,
is painted in the muted colors of
the cactus and succulents.

Lester Hawkins was called in
to lay out a scheme
of curving paths and beds
at the center of the two-acre site.
Two screened-in pavilions
at the entrance of the garden
provide shelter for plants
too delicate to survive
the wintertime low temperatures
of twenty-six
to twenty-eight degrees.
A connecting pavilion
and greenhouse were added later.

bulbs, people began to stop by each spring to admire the flowers. Eventually, the family established a routine of opening the ranch to the public for the weeks of peak bloom, usually from mid-March to mid-April.

Today, their descendants continue the tradition, planting up to four thousand new bulbs each year. As many as twenty thousand visitors come to wander through the meadows spangled with an estimated four hundred thousand flowers.

Mary Lucot Ryan, granddaughter of Arthur and Lizzie McLaughlin, confesses that no one has ever really counted. "Uncle Jesse, who died in 1981, used to say there were three hundred thousand, but I have no idea where he got that figure. We say four hundred thousand now, but frankly, no one has the slightest idea."

She says that about four hundred varieties of daffodils, tulips, grape hyacinths, jonquils, narcissus, and other bulbs thrive and multiply at Daffodil Hill. To add to the charm, the fields are dotted with flowering cherry, apple, plum, and crabapple trees. The massive lilac bushes that flourish near the old cabin were already here when the McLaughlins bought the property. The barn, the same one that used to shelter travelers' horses in the 1880s, and other farm buildings of log construction make it easy for visitors to entertain fantasies of rural life a century ago.

Low-energy gardeners will be cheered to know that this lovely scene is low-maintenance gardening at its best. Arthur Lucot, Mary Ryan's brother, orders the bulbs, specifying single-blossom daffodils rather than the more expensive double-bloom varieties. Thanks to excellent soil, winter snows, and long, dry summers, the bulbs take care of themselves. Family members plant new bulbs and, once in a while, divide plants in crowded areas. In 1989, at the urging of a daffodil expert, they broadcast bone meal across the fields—the first time that the garden had ever been fertilized.

Children, grandchildren, and friends do most of the work at Daffodil Hill, starting with the planting of bulbs in late fall and finishing with weeding in the spring. During the weeks that the ranch is open to the public, as many as six McLaughlin descendants help direct traffic. A local youth group operates a snack bar just across the road.

They have never advertised, other than to let newspapers know their opening and closing dates, nor have they charged admission, although they're happy if visitors want to put donations in the yellow teapots hung near the barn.

Mary Ryan says, "My family wanted a place where families could come with their children without spending a fortune. People keep suggesting that we carry T-shirts and sweatshirts and whatever, but the only thing we sell is postcards—at the same price Uncle Jesse charged in the 1950s."

Visitors respond to their generosity with a high degree of consideration. A state park official once asked Mary Ryan how big her clean-up crew was. "What clean-up crew?" she responded. "People don't leave a mess here, even in the picnic area."

Daffodil Hill is a private, family-run operation. By planting walnut groves on the 540-acre ranch, Arthur Lucot and Mary Lucot Ryan hope to have created a financial support to enable their descendants to carry on the tradition. Although this field is only fifty miles from the cities of Sacramento and Stockton, it offers an incomparable chance to see a world only lightly touched by the twentieth century.

❦ DAFFODIL HILL

18310 Ram's Horn Grade Volcano, CA 95689 209/296-7048. Daily from mid-March to mid-April only, 10 to 5. Closed in rainy weather. Donations appreciated. Take Highway 88 east through Jackson, crossing Highway 49 in the Gold Country, to Pine Grove; follow signs to Volcano and Daffodil Hill.

The McLoughlin family has
resolutely avoided
doing anything
that would make
Daffodil Hill more commercial
or less authentic.
As always, they plant bulbs
in whatever container comes to
hand—kitchen sinks, lengths
of terra-cotta pipe,
washing machine drums,
a steam table,
a stove, or an old pot.
Somehow,
the unstudied approach works,
creating a cheerful chaos
of blooms everywhere you look.

Families flock
to Daffodil Hill each spring
to delight in the abundance
of some four hundred varieties
of daffodils, tulips,
hyacinths, jonquils, and narcissi,
bursting forth from fields
shaded by flowering cherry,
apple, plum,
and crabapple trees.
Old farm equipment
and a barn that once served
as a stopping place
for travelers crossing the Sierra
in the 1880s
give a vivid sense
of California life a century ago.

The link between spiritual, ecological, and practical concerns gives Green Gulch a special serenity, often felt by visitors as soon as they turn off Highway 1 to descend the winding, eucalyptus-lined driveway. Visitors have a chance to appreciate the beauty and calm of the valley as they walk along a dirt road from the parking lot towards the main gardens.

A high wall of cypress trees marks the eastern edge of the garden, separating it from nearby buildings and creating a sense of protected space.

Head gardener Wendy Johnson Rudnick explains, "The garden was laid out with contemplation and beauty in mind. One third of it is spaces for people to sit. The hedges give a sense of closure and privacy."

The one-and-a-half-acre plot has been divided into smaller sections, each designed to show off plants and to provide quiet spots for meditation. The quietest, most protected part of the garden, tucked up against the hillside, was deliberately planted in a restrained color scheme of green and white. Certain plants have special significance to Buddhists. Wendy Rudnick points out a grouping called the three friends—pine, symbolizing strength, bamboo for flexibility, cherry for beauty and transitoriness. Some of the trees have been planted by Zen Center members as a gesture of remembrance for friends and relatives who have died.

The small, peak-roofed shrines that dot the garden honor Jizo, the patrol saint of travelers and children. The layered red aprons are put on the saint by mothers participating in a ceremony dedicated to children who have died, particularly those lost to abortion, stillbirth, or miscarriage.

The garden is unusual for more than its spiritual qualities. Its design incorporates flowers, shrubs, fruit trees, and herbs in close proximity, showing the influence of Alan Chadwick, a legendary California gardener whose French intensive, biodynamic gardening methods emphasize deep digging and close planting of compatible plants. Chadwick helped develop Green Gulch and lived here for the last few months of his life.

The one-and-a-half-acre garden produces such an abundance of riches that its flowers, fruits, and herbs are sold to Bay Area markets and restaurants, including Greens, the Zen Center's own vegetarian restaurant on San Francisco's waterfront. Members of the public are also encouraged to buy plants here throughout the year, especially at the annual spring plant sale in late April.

Beyond the nursery and the greenhouses full of seedlings, the road leads towards the Pacific Ocean through wide fields of vegetables. Peter Rudnick, Wendy's husband, is the manager of Green Gulch's farming operation. As such, he's one of the people who have transformed the cuisine of California, both in restaurants and at home.

Green Gulch Farm's success at growing and marketing organic vegetables has made it one of the most effective Davids to stand up against the Goliath of California agribusiness. Thanks to Green Gulch and other so-called boutique growers, many Californians now differentiate between potatoes called Ruby Crescent, Yukon Gold, and Yellow Bintje, choose among myriad salad greens, and sprinkle them with edible flowers. The Green Gulch influence is obvious even in the produce departments of large supermarkets, which now offer choices no one even knew about a decade or so ago.

If a beautiful garden, a spiritual life, unusual plants for sale, interesting gardening methods, and stimulating classes aren't quite enough to tempt you to visit Green Gulch, maybe the masses of summer flowers will. As a friend said, "Go, if only for the sweet peas."

❧ GREEN GULCH FARM
1601 Shoreline Highway Sausalito, CA 94965 415/383-3134. Garden tours are by reservation, or drop by on Sunday, the public day. Donation of $5 requested. From Highway 101 north of Sausalito, take the exit for Highway 1 and Stinson Beach and continue to the stoplight (Tamalpais Junction); turn left to stay on Shoreline Highway. In 3 miles, at the turnoff for Muir Woods, keep left. The farm is 2 miles farther on the left, just before Muir Beach. Follow signs to parking lot and the gardens.

Fields of organic vegetables
help to support
the spiritual activities
of Green Gulch Farm's
residential Zen community.
Elsewhere in the garden,
the proximity
of flowers, shrubs, fruit trees,
and herbs
shows the influence
of Alan Chadwick,
a legendary California gardener
who emphasized
deep digging and close planting
of compatible plants.

SONOMA HORTICULTURAL NURSERY

Sebastopol, California

The cheerful music of canaries fills the air at Sonoma Horticultural Nursery. Their song is a subtle introduction to Polo de Lorenzo, a native of the Canary Islands who is one of the owners of this fine specialty nursery and its acres of gardens. Sonoma Horticultural Nursery, on a sloping hillside in the farming country of western Sonoma County, is a mecca for rhododendron fanciers. They typically have a choice of about three hundred species and three hundred hybrids here. After a normal winter, when temperatures here usually drop to about nineteen degrees, the rhododendrons are in full bloom in late April and early May.

Sizes range from five-gallon tubs, lined up by the hundreds in the shaded nursery area, to full-grown specimens planted in the garden. They're all for sale, even the biggest ones. The staff is particularly proud of their collection of *Madden* rhododendrons and of their dwarf varieties, some of which grow only knee high and are so small that people mistake them for azaleas.

Azaleas, closely related to rhododendrons, are also well represented here, with some two hundred varieties for sale.

Polo de Lorenzo doesn't stop with these two huge groups of plants, as a walk through the six-acre garden proves. He is an enthusiastic experimenter, using the garden to provide new impetus for the nursery operation and vice versa. Things are in a continual state of flux here, as the garden is expanded and enriched, with new pleasures for the eye added every year.

After parking in the lot at the bottom of the hill, visitors should resist the temptation to head directly into the greenhouses and shaded patches of the selling area. Instead, they should start a visit to the garden a few yards up the hill, at the more formal entry near the house. Two circular areas show off some of the most beautiful azaleas and rhododendrons, some growing in billowing mounds at the foot of tall pine trees. Masses of deep purple columbine, roses, magnolias, and clematis vines with huge white blossoms fill in the gaps at different times of year.

The davidia tree next to the house is extremely rare, so much so that this particular specimen has been designated a Sonoma County Heritage Tree. Its huge blooms, which start as pale chartreuse and become a pure matte white, appear in April.

From the driveway, a narrow dirt pathway winds under the Monterey pines that border the upper edge of the property, providing shade and wind protection for some of the garden's most impressive rhododendron specimens. In March and April, the explosion of blooms forms a wide ribbon of color through the dappled shade. Specimens here range from the soccer ball–sized pink-and-white blooms of the variety Diane Titcomb to the delicate pale pink Mrs. Charles Pearson and the unusual coral-tinged, pale yellow Golden Pheasant.

The feeling of a natural woodland is intensified by the interesting variety of companion plants interspersed among the azaleas and rhododendrons. The sharp-eyed will find the unusual *Primula vialii*, with its two-foot-high flower spikes, and lady's slipper orchids along the way. Ferns, dicentras, and hostas also thrive in this shady area. They're combined most dramatically around a small picnic table that offers a view of the large pond's rim in the center of the garden.

The pond once served as a watering hole for the cows that occupied this property before de Lorenzo and his partner, Warren Smith, took it over in 1964. Today, the pond's rim is planted so that waves of color appear throughout Northern California's long blooming season. In spring, deep blue Japanese iris, azaleas, rhododendrons, and wisteria provide the color. In summer, daylilies offer bursts of warm yellows and oranges. Other interesting plants around the pond include candelabra primulas and some unusual flowering fruit trees, including cherries and crabapples. The edges of the pond are planted with pre-historic looking massive gunnera.

More of Polo de Lorenzo's continuing experiments are beginning to fill up the boggy area at the lower edge of the property, which overlooks a neighboring llama farm. One of the most successful is the yellow garden, inspired by some of his favorite English gardens. *Corylopsis spicata, Corylopsis pauciflora*, yarrow, and variegated green-and-yellow boxwood contribute to the sunny effect. Although the glow lasts in one form or another all year, this part of the garden is at its best in spring and summer, after the rhododendrons and azaleas have faded.

Elsewhere, an allée of Japanese maples is sparked by companion plantings of yellow daylilies and species of geraniums. Near the yellow garden, unusual conifers are planted in circular, stone-walled beds. Each small area is linked by paths of paving stones interspersed with a soft carpet of chamomile, an attractive ground cover with an appealing smell.

If there is a single motif uniting all the disparate parts of this garden, it is a generosity of spirit. Nothing is done in a half-hearted or self-consciously delicate way. Marguerites are planted in huge billows, and eight-foot-high hollyhocks grow in such abundance that they form a forest. De Lorenzo's enthusiasm is contagious, to the people who visit here and, apparently, to the plants themselves.

❧ SONOMA HORTICULTURAL NURSERY
3970 Azalea Avenue Sebastopol, CA 95472 707/823-6832. Daily, Thursday to Monday, 9 to 5. Free. From Highway 101 north of Petaluma, take the Highway 116 exit west to Hessel Road; turn left, bear right where Hessel meets Turner Road, then turn left on McFarlane Road and right on Azalea. Visit in April for rhododendron and azalea bloom.

The central axis
of the English-inspired
yellow garden
is a long-trellised archway (above)
covered with golden
laburnium blossoms,
leading the eye
to a small terra-cotta fountain
at the far end.

A pond (top left),
once a watering hole for cattle,
is the centerpiece
of Sonoma Horticultural
Nursery's garden.

Other civilized touches
in this constantly
evolving garden
include fountains, benches,
and walkways
of stones laid
in a carpet
of fragrant chamomile.

The verdant banks of the pond
are usually radiant with flowers,
starting with azaleas and
rhododendrons
in the early spring, irises in late
spring, and daylilies
during the summer months.
A wooden structure is home
to dozens of canaries
whose song adds to the charm
of this Sonoma county nursery.

to best effect. Their goal was to see just how many temperate and subtropical species they could grow on this site a few miles inland from the Pacific Ocean. Their property receives an extraordinary amount of rain, about sixty inches a year, followed by a six-month stretch when there is dampening fog but no rain at all. In recent years, California's long drought, starting in 1987, and an unusually long period of subfreezing temperatures in December 1990 have added to the challenge.

Over the years, Western Hills became a kind of informal horticultural salon, a place where horticulturists and gardeners gathered to exchange information and examine the rarities that Olbrich was continually receiving from friends and colleagues around the world. The conversation was not limited to botanical matters; Olbrich was a fiercely opinionated intellectual who was interested in history, philosophy, books, and the arts.

Their influence spread in other ways. They were active in state horticultural associations, often introducing their new plant discoveries at club meetings. In addition, they worked on a number of important gardens, contributing both design and plants.

Their own garden developed gradually, in pace with the owners' varying energies and finances. They never had a formal plan for their undulating slope, which in any case was unsuited to formal design conventions such as a central axis, terraces, or obvious symmetry. Instead, they kept planting new things in a spirit of experimentation disciplined by a strong aesthetic sense. Western Hills' soft-edged, cottage-style plantings are exceptional because of the subtle sense of color and discerning eye applied to the combinations of heights, shapes, and textures.

Olbrich wrote, in Rosemary Verey's *The American Man's Garden*, "Our garden seems to be more like a rather disorganized novel or drama in which characters come and go, with the good pining away and the wicked sometimes flourishing."

Whether they are the good guys or the bad guys, Western Hills' plants always seem to belong together, at least for a while. Among the hundreds of vignettes, memorable sights include the view down the slope in front of the house. Olbrich and Hawkins created a landscape that is almost Japanese in its emphasis on texture and form, using plants with silvery and gray-green foliage that form spiky and grassy mounds. The components include xanthorrhoea, euphorbia, phormium, helictotrichon, fescues, and a weeping silver pear tree.

Elsewhere, the eye is caught by color. A summer border combines the soft blue of love-in-a-mist (*Nigella damascena*) with the brilliant orange of California poppies and the purple flowers and gray foliage of lavender. Near the picnic table, linarias, roses, clear blue geraniums, and dogwood create a delicate setting in shades of pink, blue, lavender, and red.

Everywhere, gardeners will find inspiration in the layering of plants of different heights. Hollyhocks, for instance, tower five or six feet tall over a carpet of violets. One of Olbrich's many interests was ground covers, so that almost every bit of open space is covered with something low, green, and often flowering. As his own garden grew and became shadier, he took a special interest in low-light ground covers, such as *Geranium macrorrhizum* and *Symphytum grandiflorum*.

Although its original owners have died, Western Hills continues to operate, inspiring gardeners far and near.

❧ WESTERN HILLS NURSERY
16250 Coleman Valley Road Occidental, CA 95465 707/874-3731. April through October, Wednesday through Sunday, 10 to 5. Free. From Highway 101 north of Petaluma, take the Highway 116 exit west to Highway 12 (Bodega Road). At the crossroads before Freestone, bear right onto Bohemian Highway, continuing to Occidental. Turn left onto Coleman Valley Road.

Western Hills'
original owners,
Marshall Olbrich
and Lester Hawkins,
used the garden as
a laboratory for new species
and new designs.
Its informal layout was
conducive to the creation
of hundreds
of small vignettes
to inspire
California gardeners.

UNIVERSITY OF CALIFORNIA DAVIS ARBORETUM

Davis, California

In 1936, a botany professor at the University of California at Davis gazed out at the endless fields surrounding the campus and decided that his students needed a greater variety of plants to study. Thus was the UC Davis Arboretum born. The first trees were planted on both sides of Putah Creek. Since then, the arboretum has continued to grow, both in numbers of species and in acreage. Some of California's finest landscape architects, including Lawrence Halprin, Ted Osmundson, and Mai Arbegast, have contributed to its design.

Botany students come here to study and do research, along with entomologists, hydrologists, and art students, whose drawing classes often take place among the trees. On baking hot summer days, one of the coolest places in town is the arboretum's grove of coastal redwoods.

Unlike other arboreta in Northern California, UC Davis' does not benefit from the cooling influence of summer fog. A plant that survives here not only must resist drought and frost but must be extremely heat tolerant as well. The average winter low in this part of the Central Valley is a fairly benign twenty-eight degrees, but summer weather is more extreme. The average high on a summer day is ninety-five degrees, with the thermometer occasionally soaring up to one hundred twenty degrees. Even in normal winters, water is scarce, rainfall averaging about seventeen inches annually.

species that will do well in these conditions. By the early 1990s, an estimated thirty-seven thousand different plants were thriving here, with an additional thousand being added each year.

As the name indicates, the main focus here is trees. At the eastern end of this long, skinny garden, a grove of eucalyptus displays some seventy species of this Australian tree, now so well adapted to California. The redwood grove is the largest display of *Sequoia sempervirens* away from its usual coastal habitat; the tall trees, it seems, will flourish away from the fog as long as there are enough of them to create a microclimate.

Moving west towards the center section of the garden, one comes to collections of native and exotic conifers and of acacias. Unfortunately, the latter suffered considerable damage in the exceptionally long stretch of cold weather that hit Northern California in December 1990.

The central section, between California Avenue and the Weier Redwood Grove, displays the arboretum's collection of California native plants. In spring, this is one of the best places to appreciate the brilliant colors of California's shrubs, perennials, and annuals. Filling out the spring palette are the bright magenta flowers of the western redbud (*Cercis occidentalis*), the gaudy orange of California poppies, the lavender and yellow spikes of lupines, and showiest of all, the golden yellow flowers of the flannel bush (*Fremontodendron* 'California Glory').

Somewhat more subtle effects occur in the Peter J. Shields Oak Grove at the western end of the arboretum. This is the largest collection of oaks in the western United States, displaying some eighty-four species of oak trees plus a large number of subspecies and hybrids, scattered around a large lawn that's a popular recreation area for students.

A half-hour stroll along the pathway is a pleasant, shady education in the surprising variety of oaks that grow around the world. Signs call attention to some of the most interesting species. Appropriately for a university that houses one of the world's leading viticulture and winemaking schools, one of these is the cork oak, whose bark is still the best way to seal a wine bottle.

Adjoining the oak grove are two small areas designed to inspire the home gardener in hot, dry locations. The Ruth Risdon Storer Garden is divided into three areas receiving various amounts of water. The shadiest section under the oaks needs regular summer watering, while the driest section can be watered once a month in summer. The planting consists of both familiar natives and exotics and some unfamiliar species from other dry areas of the world.

The garden seems much larger than it is because of its dense planting and convoluted pathways. It was named after a woman with a special talent for creating harmonious and subtle color combinations in her own gardens. In tribute to her, many of the plants in the garden have soft gray or gray-green foliage and flowers in many shades of lavender and pink.

A few feet away, the Carolee Shields Garden surrounds a large gazebo. Here, water conservation is less important than aesthetics. This is a white garden, abloom with dozens of white-flowering plants that shimmer in the moonlight as well as the sun. Successive bloom comes from narcissus, star-of-Bethlehem, and verbena. To give the garden a silvery look at night, some of the foliage plants have silver or gray leaves; these include wooly lamb's ears (*Stachys byzantina*), dusty miller (*Senecio cineraria*), and blue fescue, a bunch grass that grows in small clumps.

Before touring, stop at the garden's headquarters to pick up some of the informative leaflets available there. With these in hand, the UC Davis Arboretum is an education as well as a pleasure.

❧ UNIVERSITY OF CALIFORNIA DAVIS ARBORETUM
University of California at Davis Davis, CA 95616 916/752-2498. Daily, 24 hours. September to June, tours on Sundays; call for scheduled times. Free. From Highway 80 west of Sacramento, take the UC Davis exit, following Old Davis Road to the campus. Across the bridge, turn left on La Rue Road; the arboretum office is on the left.

Students and Davis residents
enjoy a changing
palette of color
as they walk and jog along
the pathways
and bridges
of the University of California's
arboretum at Davis,
which specializes
in plants able to tolerate
both drought and heat.

Putah Creek, running along
the southern edge of the campus,
established the long,
sinuous shape
of the arboretum at Davis.
Hundreds of species of trees,
both native and exotic,
cast their reflections
in the water and create
a welcome ribbon of coolness.

MENDOCINO COAST BOTANICAL GARDENS

Fort Bragg, California

rom a sailor's point of view, the Mendocino coast, with its rocky headlands and crashing surf, is not a hospitable place. From a gardener's standpoint, however, this rough coastline offers happy surprises. Small enclaves of subtropical climate create ideal growing conditions, and the sailor's dreaded fog is the gardener's summer delight.

Ernest and Betty Schoefer began looking for one of these "banana belts" when he retired from his nursery business in Southern California. In 1961, they bought a forty-seven-acre site three miles south of the lumber town of Fort Bragg. The property, sloping gently from Highway 1 to the Pacific, was perfect for a wilderness garden of both native and exotic plants.

For sixteen years, the Schoefers cleared the land of brush and fallen trees, got rid of derelict buildings, cut trails down to the sea, and planted rhododendrons, azaleas, fuchsias, and ferns. Gradually they created the Mendocino Coast Botanical Gardens, emphasizing lushly naturalistic plantings rather than encyclopedic displays of different species.

In 1977, the Schoefers sold their acreage to private investors, who in turn sold twelve acres—the garden itself—to the California Coastal Commission. A nonprofit corporation took over the operation of the garden, nursing it along without any means of financial support and with the constant threat of nearby development. In 1986 the investors' plan to develop the surrounding property was withdrawn after public furor. In 1991, after suffering financial difficulties, the owners sold the remaining thirty-five acres, so that the original forty-seven acres are now under the aegis of the foundation.

In similar circumstances, most gardens would simply have disappeared. Fortunately, Mendocino was blessed not only with a gentle climate but with active, knowledgeable volunteers, who fought for years to ensure that the Schoefers' efforts would not be lost. Dr. Leonard Charvet, a retired surgeon who has been one of the garden's most active supporters, said proudly, "This is a people's park."

Today, a small paid staff operates the garden with the help of hundreds of volunteers. The many local nurserypeople have provided plant collections and advice. Individuals propagate and hybridize plants at home for the botanical garden. Local timberpeople have done some of the dangerous work involved with clearing trees. A man in his eighties built all the signs and more than a hundred benches, while youthful members of the California Conservation Corps have done much of the hard physical work. The fruit of their labors is a spectacularly beautiful garden that takes seriously its role as a repository of rare and fragile plants.

The garden is divided into three distinct sections. Display gardens, attractively planted in circular beds separated by lawn, occupy the sunny, level area near the entrance. Collections include old-fashioned roses, some of which are the hardy climbing varieties one sees on Mendocino County farmhouses. Other beds hold camellias, succulents, dwarf conifers, and a colorful selection of perennials. Considerable space is given to heaths and heathers and to dwarf and tender species of rhododendrons. Nearby, a handsome lath building houses 350 varieties of ivy, donated by the American Ivy Society in 1986. In May, a local fuchsia specialist brings in his collection to add color to the ivy display.

From this level area, several trails wind down through a verdant canyon, where streams bubble over waterfalls and a high canopy of pine trees provides dappled shade. Native ferns, large-leaf rhododendrons, azaleas, and magnolias grow here in rain-forest lushness; Mendocino Coast Botanical Gardens boasts some thirty thousand examples of thirty different varieties of rhododendron.

As visitors proceed along the winding trails (most of which are accessible to people in wheelchairs), they will note the small pocket gardens that are one of Mendocino's most charming details. These small, informal clearings show off smaller plants, including Pacific iris, daffodils, and dahlias.

A barrier of Bishop and shore pines not only protects the canyon from wind but draws a line between the dimness of the forest and the brilliant light of the bluffs along the ocean. These windswept meadows are splashed with iceplant, flowering in bright pink or yellow, the bright orange blooms of the poker plant (*Kniphofia uvaria*), marguerites, and other wildflowers. Even the most dedicated horticulturist will be distracted from the plants by the view of craggy rocks and roiling surf. The sharp-eyed may spot sea lions and whales at certain seasons.

As beautiful as the Mendocino Botanical Gardens are now, they seem destined to become even more interesting in the years to come. In the first months after receiving the additional thirty-five acres in 1991, Mendocino's energetic staff and volunteers planted a wildflower meadow, pruned an old orchard, and refurbished a little house on the cliffs to give visitors protection from the wind. Other projects are under discussion. As Dr. Charvet phrased it, "This is not a finished garden. This is a starting garden."

❧ MENDOCINO COAST BOTANICAL GARDENS
Mail: P.O.B 1143 Location: 18220 North Highway 1 Fort Bragg, CA 95437 707/964-4352. March to October, daily, 9 to 5. November to February, daily, 9 to 4. Adults, $5; seniors, $4; students, $3; ages 12 and under, free. On Highway 1, 6 miles north of Mendocino and 2 miles south of Fort Bragg. Visit in April and May for rhododendron bloom.

A wide variety of plants
flourish in a tiny
microclimate
on the rough
Mendocino coast.
In the more formal
demonstration gardens,
heaths and heathers
are among
the featured attractions,
while sturdy iceplant,
marguerites,
and poker plants
brighten the cliff area.

The burble of streams
and the dappled shade
of pine trees
create a romantic woodland that is
a perfect setting for ferns,
rhododendrons,
and azaleas, interspersed
with tiny pocket gardens
of other woodland plants.

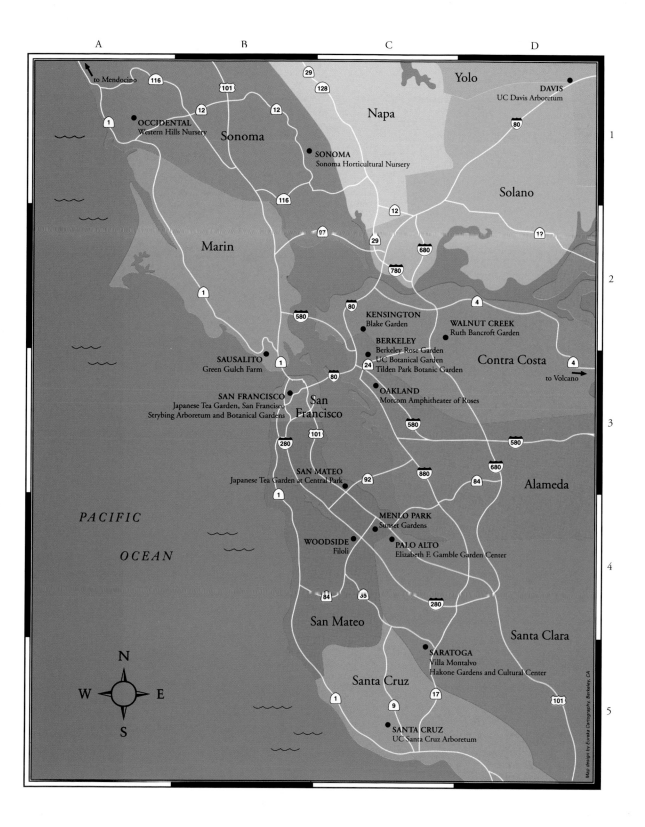

A
B
C
D

to Mendocino

116

101

128

29

Yolo

DAVIS
UC Davis Arboretum

OCCIDENTAL
Western Hills Nursery

12

12

Napa

80

1

1

Sonoma

SONOMA
Sonoma Horticultural Nursery

Solano

116

12

Marin

07

29

1?

780

680

2

1

580

80

KENSINGTON
Blake Garden

4

WALNUT CREEK
Ruth Bancroft Garden

BERKELEY
Berkeley Rose Garden
UC Botanical Garden
Tilden Park Botanic Garden

SAUSALITO
Green Gulch Farm

1

24

Contra Costa

4

to Volcano

80

SAN FRANCISCO
Japanese Tea Garden, San Francisco
Strybing Arboretum and Botanical Gardens

San Francisco

OAKLAND
Morcom Amphitheater of Roses

580

3

101

280

580

SAN MATEO
Japanese Tea Garden at Central Park

92

880

680

84

Alameda

PACIFIC

1

MENLO PARK
Sunset Gardens

OCEAN

WOODSIDE
Filoli

PALO ALTO
Elizabeth F. Gamble Garden Center

4

84

35

280

San Mateo

Santa Clara

N
W E
S

SARATOGA
Villa Montalvo
Hakone Gardens and Cultural Center

Santa Cruz

1

17

101

5

9

SANTA CRUZ
UC Santa Cruz Arboretum

Map design by Eureka Cartography, Berkeley, CA

1

20

MENDOCINO
Mendocino Coast Botanical Gardens

128

Mendocino

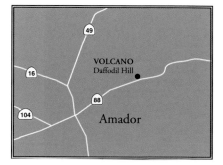

49

VOLCANO
Daffodil Hill

16

88

104

Amador